Alex Oliveira

# Lower Gávea

Alex Oliveira

# Lower Gávea

## Outstanding Leisure Area in the City of Rio de Janeiro

ScienciaScripts

Cover image: www.ingimage.com

This book is a translation from the original published under ISBN 978-3-330-75589-5.

Publisher:
Sciencia Scripts
is a trademark of
Dodo Books Indian Ocean Ltd. and OmniScriptum S.R.L publishing group

120 High Road, East Finchley, London, N2 9ED, United Kingdom
Str. Armeneasca 28/1, office 1, Chisinau MD-2012, Republic of Moldova, Europe
Printed at: see last page
**ISBN: 978-620-8-05648-3**

# INDICE

# Dedication

To my mothers, Eunice Silva Nogueira, Sonia Rodrigues and Tania Rodrigues, for their love, attention, help and affection.

# CHAPTER 1 - INTRODUCING THE WORK

When we think of the many different kinds of events that exist in the world today, geographical content is not always correctly remembered or inferred. Everyday life and its events have always been the main focus of scientific studies. There are various popular terms, ideas and notions which, despite being highly widespread, have not received any kind of discussion in the human sciences, including geography. Reflecting on the spatio-temporalities of the contemporary world is a task given to the sciences in general, so Geography, by trying to understand the spatial organization of certain times, is also interested in the scientificization of these "public knowledge" themes.

When we hear expressions like "let's go to the bar", "let's have a beer", "let's have an ice" and the like, we can't help but think of a gathering of friends chatting and having fun around a bar table. The tone of "pleasantness" is associated with the idea by all possible audiences, including teetotalers, whether it's because of the acclaimed bohemia of the old carioca malandros or simply because it's a moment of fun between friends. The idea of getting together in a bar is much more than just enjoying a few alcoholic drinks.

One of the activities that cariocas do most at night is go to their favorite bars and nightclubs. We don't realize it, but when a certain area of a neighborhood has the busiest nightclubs, it is nicknamed "baixo do bairro". All the *bars* scattered around the marvelous city are so named because of the influence of the pioneer and most famous of them all, *Baixo Gavea*. Geography, despite dealing with cross-cutting themes similar to the phenomenon of the Baixos, has never been very concerned with their specificity. What we will try to do is analyze, through a geographical lens, one of the most famous bohemian spots in the city of Rio de Janeiro.

Located in Praga Santos Dumont, in the Gavea neighborhood, *Baixo Gavea*'s main characteristic is the profusion of services related to leisure and entertainment in Rio, which attract different audiences regardless of the time of day. Gavea is a residential neighborhood in the south of the city of Rio de Janeiro, where most of the population is considered to be middle-class or upper-middle-class. The neighborhood's history includes the refunctionalization of its space, due to the rise in value of real estate in the south zone, because at the beginning of the last century, Gavea was considered a proletarian neighborhood because it housed workers from the factories that were located near or inside it. The area in question has a global resonance, having already been the subject of novels, the backdrop of biographies, referenced in songs and the meeting place of artistic celebrities in Rio de Janeiro.

The interest in the subject of this work lies in the enthusiasm to add perspectives to urban geographic studies, believing that common sense should be even more appropriated and studied by geographic science. At the same time, we can't escape the personal interest that arose after spending so much time in Praga Santos Dumont and various other low-rise neighborhoods, and noticing that there was no scientific study or definition (albeit incipient) for the name of a "low-rise neighborhood".

In view of the above, this work aims to: (a) take a geographical look at *Baixo Gavea*; (b) trace the urban evolution of this research object; and (c) investigate the concepts that will serve as the basis for our research. Thus, our research will bring us the attempt to contribute in a relevant way to science as a whole and, specifically, to future research that relates to Gavea, the South Zone or the city of Rio de Janeiro.

In this way, the work seeks to analyze and discuss "how could the low be a geographical category?". This question is due to the fact that the issue in question has not been widely discussed or researched in other geographical works. We intend to try to resolve this question by presenting plausible arguments that demonstrate the geography of this street. A subsequent question would be "what is the importance of the basses, especially Gavea, to the neighborhoods where they belong?". With these questions, we intend to try to address the factors that make this agglomeration of night-time services so characteristic and important for the city. In addition, the research process revealed an interest in knowing "what other functions might *Baixo Gavea* have?".

Our work will be based on bibliographical contributions relating to Geographical Space, Urban Geography and similar subjects; we will also use secondary sources obtained from government agencies such as IBGE, IPP and the National Library, as well as using material available on the internet.

We will use the sources to complement our research, following the geographic reasoning of the urban current, because its theoretical scope is geared towards thinking about the city and its nuances. Our aim was to try to interpret the questions that arose with a geographical eye. We would like to point out that, as it is not a topic widely discussed in academia, and is considerably complex for an undergraduate monograph, we do not have an extensive reference framework.

The book will present four more chapters dealing respectively with the following subjects: definition of the methodological bases for the analysis of the object; urban evolution of the Gavea neighborhood, demonstrating all its dynamics; spatial analysis of *Baixo Gavea*; and final considerations about the research

# CHAPTER 2 - GEOGRAPHICAL SPACE: A BASIC CONCEPT FOR SCIENCE

Geographical space has been referred to by many as the primary object of our science. Its importance is not confined to Geography, but is the subject of discussion in other sciences. This concept, alluded to as one of the five fundamental concepts of Geography (in addition to the concepts of Region, Landscape, Territory and Place), and its conception, are intertwined with the history of geographical thought itself. The discussion of this concept is so extensive that geographers such as David Harvey, Horacio Capel, Roberto Lobato Correa, Milton Santos, among others, have dedicated important works to enriching the reading and research of other geographers.

In this chapter, we will build the conceptual scope of the monograph, going through and discussing the concepts of space up to the critical current. Thus, we will talk about the concept of space from an urban perspective, some important categories for our subject, the issue of leisure and leisure in urban space, as well as the nuances surrounding public space.

Space is commonly associated (and sometimes taught) as a specific portion of the earth's surface. This does not take into account the complexity of the concept, given that it is not related to society, which is also an object of study in geography, just like the earth's surface. In order to discuss and use the geographical space in a broad way in our research, we will then go through the conjuncture and formation of the concept in question through the different geographical currents, up to the Critical current that will best serve our object of study.

Traditional geography (circa 1870 - 1950), although its key concepts were region and landscape, began discussing geographical space based on the considerations of Ratzel, with the concept of vital space, and Hartshorne, with the idea of absolute space, derived from the "Lablachian" concept of region.

The vital space, conceived by Ratzel, is closely linked to German government interests, although it shouldn't be totally dismissed by science, after all, the concept was a precursor to a spatial reading that was more connected to society. In the words of Moraes (2003), the vital space would be,

> a proportion of balance between the population of a given society and the resources available to meet its needs, thus defining its potential for progress and its territorial priorities. [...](the vital gap) is expressed in the justification of expansionism as something natural and inevitable in a society that is progressing, generating a theory that legitimizes Bismackian imperialism (p. 18).

For Hartshorne, geographical space has no interaction or interference with human support. Hartshorne's writings are based on Vidal de La Blache and the French school of geography. This influence can be seen in the intrinsic link between Absolute Space and the concept of Region, just as Ratzel's Vital Space is associated with the concept of Territory. La Blache's Region was a concrete entity that existed on its own, and so Hartshorne defined space as "a receptacle that merely contains things" (CORREA, 2008, p. 18-19). Correa also gives us a quote from the author in question in which space is used merely as an area, since

> it is only an intellectual picture of the phenomenon, an abstract concept that does not exist in reality (...) the area itself is related to the phenomena within it, only in that it contains them in such and such locations (HARTSHORNE apud CORREA, 2008, p. 19).

In theoretical-quantitative geography, hypothetical-deductive reasoning combined with logical positivism brought major changes to geography and its objects of study. The concept of space appeared as the key to the science, leaving aside the concepts of Landscape, Place and Territory Two typologies emerged for the concept of space: the matrix representation and the isotropic plain.

The matrix representation, as Correa (2008) points out, is widely used by spatial economists, based on the ideas of John Nystuen[1] , bringing to Geography "a limited view of space, since [...] distance is privileged too much" (CORREA, 2008, p. 22), not considering other important factors in the construction of space.

The isotropic plain, on the other hand, would be a uniform space in terms of geomorphology, vegetation cover, climate and human occupation, bringing a sense of equality to all the forms and functions exercised in that locality. In this concept, distance brings spatial differentiation, which would generate a balance between the different plains[2] .

Parallel to the isotropic plain is the notion of relative space[3] , which can be understood by the relationships between objects, which would be essential to override distance, which until then was the category/concept considered basic for geographical studies, along with orientation and connection. The relativity of space leads us to think about the relational factor of objects and functions in space, although in a context of theoretical geography, David Harvey initiated spatial thinking beyond its form, a fact that is more noticeable from the 1970s onwards.

Bessa (2010) presents us with a consideration of how the theme of spatial differentiation, so dear to geographers today, was developed during theoretical geography, when he states that

> the resulting spatial patterns are based, because of the assumptions of economic rationality, perfect competition and ahistoricity, on a situation of spatial uniformity, expressed by means of an isotropic plain, which in principle disregards heterogeneity, but on this isomorphic space, economic supports and mechanisms are developed which lead to the diversity of the space (p. 47).

The development of the element of spatial differentiation and the concept of relative space in the light of historical materialism and dialectics

---

[1] American geographer who wrote the article *Identification of some fundamental spatial concepts,* one of the greatest contributions to theoretical geography.

[2] Theories such as Weber's industrial localization, von Thunen's concentric rings and center-periphery schemes were derived from this concept.

[3] Brought to us by geographer David Harvey, in 1969, with his *explanation in geography.*

(already incorporated at the time by other human sciences) helped geographers to change the paradigm of Geography itself. These approaches broke with the ideas and concepts previously mentioned in Geography, giving rise to the Critical current of science. From then on, space was analyzed in many different ways, providing geographers with countless interpretative possibilities.

We can't talk about space in Critical Geography, or even about Critical Geography itself, without mentioning the geographer Milton Santos, one of the precursors of this critical current in Brazil. Based on foreign authors, he constructed the notion of space, which has been widely disseminated in Brazilian geography to this day. In addition to the recognized contribution of various Brazilian geographers to the thoughts and concepts that emerged in the post-1970s, we will briefly focus on Milton Santos' *The Nature of Space* (1996), where the author discusses the constructions and elements present in geographical space, as well as its relationship with time and the application of technology.

In this book, Santos presents us with a geographical space that establishes a system of objects and a system of supports that interact with each other. The systems of objects encompass "the productions of social elaboration" (op. cit., p. 64), which have their own existence and will generate functional complexity (of use) and structural complexity (of function), from the supports throughout their history. Santos adds to the importance of the relational factor of objects in recognizing the production process in general, when he says that by recognizing objects in the landscape and in space, we are alerted to the **relationships** that exist between places. These **relationships** are responses to the productive process in the broadest sense, ranging from the production of goods to symbolic production (SANTOS, 1996, p. 71, emphasis added).

Milton Santos, when talking about the system of supports, adds the factor of intentionality, but does not dissociate the supports from the objects. In this way, the supports, which correspond to the flows, would have as their driving element the intentionality of the acts, which give function to the objects (corresponding to the fixed ones) and converge, as a whole, towards a geographical object. According to Santos, "only man has an action, because only he has an objective, a purpose. [...] the purposes related to actions are carried out through individuals" (1996, p. 82), which emphasizes the role of man as a geographical agent of action.

The intentionality of support systems, presented by Santos, brings us back to the issue of spatial practices, addressed by Correa (2008), which are "a set of spatially located actions that have a direct impact on space, altering it in whole or in part or preserving it in its forms and spatial interactions" (p. 35). These spatial practices will be indispensable for achieving the spatial differentiation that is so dear to geographical space.

Just as Santos presents us with the systems of objects (fixed) and supports (flows) of geographical space, Correa will describe the modalities that make up spatial practices, which are: spatial selectivity, spatial fragmentation-remembering, spatial anticipation, spatial marginalization and reproduction of the producing region. For our research, we'll only be looking at the practices of *spatial selectivity, spatial anticipation* and *spatial marginalization,* which will better serve to explain our object in the following chapters.

Spatial selectivity comes from the organization of man, who acts by selecting a particular location to carry out his practice, provided that this location has the necessary attributes for it. In the process of spatial anticipation, we see the definition of the location of the activity in a place, before it presents the favorable conditions for it; in this way, the space can gain characteristics and then become selective for the desired spatial activity. In spatial marginalization, we look at the differences in value that the space can receive over the years, whether it is a process of valorization or devaluation. We agree with Correa (2008) that "spatial practices [...] are not mutually exclusive: on the contrary, they can occur in combination or be complementary" (p. 36).

To conclude, we would like to move on in our geographical analysis by remembering that "different human practices establish different concepts of space" (HARVEY apud CORREA, 2008, p. 19), and these established concepts of space must be considered as "something that participates equally in the condition of the social and the physical, a mixture, a hybrid." (SANTOS, 1996, p. 19). (SANTOS, 1996, p. 86).

And what "type of space" would our object of study be? After all, what kind of space are we talking about? In this chapter, we'll try to resolve these questions by discussing the possible spaces we could have in *Baixo Gavea.*

## 1.1 The Geographical Space from an Urban Perspective

The city is an extremely important part of the social sciences, and geography could not fail to contemplate this locus of the process of production and reproduction of humanity throughout history. Correa (1989) elucidates that, for us geographers, the city will represent itself as an urban space, which can be perceived by observing the inhabitants, or by the connections that the spatial form will make with the social structure, processes and urban functions.

It is therefore necessary to explain what the urban space itself is, how it is structured, who acts in this space and what forms they take. In order to approach the concept in a satisfactory way, we will base ourselves on one of the most complete works on the subject, written by geographer Roberto Lobato Correa in 1989, *O Espaqo Urbano.* One of the ways to think of the urban space is as a "set of different land uses juxtaposed with each other" (CORREA, 1989, p. 7). These uses will end up defining the spatial organization of cities, in other words, the union of fragmented types of land use, regardless of their functions, will define the urban space as a whole.

Each part of the urban space, however independent, is linked to the others. The relationships that all the parts establish are social in nature and stem from society and its processes, and are called spatial relationships. Also according to Correa (1989), "spatial relations integrate [...] the various parts of the city, uniting them into an articulated whole" (p. 8), showing that, despite the apparent fragmentation, the urban space continues to be united by spatial relations, materialized through the various and different articulations, carried out by the flows of people, goods and support.

Urban space is also a reflection of society, of the actions that take place in the present, but also of actions that took place in the past, impregnated in geographical testimonies through spatial forms. This factor will be important for understanding our object: Santos Dummont Square was a meeting place for workers from the Gavea factories, and ended up being culturally "adopted" and valued by the neighborhood's residents over time.

When we think about the question of social reflection, urban space becomes a condition and a conditioning factor for society. Conditioned because, as long as there is the dynamics of society at its foundation, it will change at different rates and of different natures. And conditioning because of the role that spatial forms will play in the processes of production and reproduction of the relations/conditions of production and sale of goods, and of social classes and their fractions. Neighborhoods will thus be considered the sites of reproduction for the various social groups, also summing up a variable symbolic dimension for each group and the setting for social struggles.

After a complex analysis of what urban space is, Correa (1989) defines it as "a social product, the result of actions accumulated over time and engendered by agents who produce and consume space" (p. 11). But who are these agents who produce and consume space? Understanding this topic is fundamental to analyzing the motivations of those responsible for creating and maintaining this urban space.

These producers are concrete social agents, analyzed because they are responsible for the complex processes that shape (or reshape) the city. Despite shaping the urban space, social agents will not change it, according to Correa (1989),

> The complexity of the actions of social agents includes practices that lead to a constant process of spatial reorganization that takes place through the incorporation of new areas into the urban space, densification of land use, deterioration of certain areas, urban renewal, differentiated relocation of infrastructure and changes, coercive or otherwise, in the social and economic content of certain areas of the city (p. 11).

The social agents in question are: owners of the means of production; landowners; real estate developers; the state; and excluded social groups. Of these, in order to contribute to our monographic research, we will only analyze the *owners of the means of production* and *landowners*[4] , which are directly applied to our research object. It should be emphasized that the actions of these agents, in particular, will always reflect the dominant interests of those who own the land, thus serving the purpose of capitalist society.

The owners of the means of production are considered to be major consumers of space. Their search will always be for favorable conditions for production, linked to urban land and nearby infrastructure, in order to seek physical support and express the locational requirements for activities.

Factory activity is generally located in large areas far from major centers, because of the pollution and environmental disturbance it can cause to the population, as was the case in the Gavea neighborhood[5] . With the real estate speculation in the South Zone, this neighborhood was surrounded by high *status* residences, *and* it was up to the owners of the means of

[4] To find out more about the other social agents, see *The Urban Space,* written in 1989 by Roberto Lobato Correa.

[5] initially a factory because of the non-occupation of the South Zone of the city of Rio de Janeiro.

production to recognize the advantages of industrial relocation, to the detriment of other housing in the neighborhood.

Landowners have acted in an attempt to obtain the highest possible income from their properties, adding to their *status* by increasing the exchange value of the land. To add value, they exert pressure on the state to change urban zoning, or demand the construction of urban infrastructure. Depending on the value given, landowners can also become real estate developers in the sense of acting directly in land speculation.

Such social agents can create selective neighborhoods in sectors close to amenities, again citing Gavea as one of the great examples of what Correa (1989) identifies as "ex-urban peripheries", because they have added value over time and, despite being physically peripheral[6] , are not considered peripheries given the often pejorative meaning of the word.

The importance of identifying the social agents who played a role in shaping the Gavea neighborhood will shed light on the reasons why its operations were carried out. We cannot fail to mention that there was direct participation by the state, real estate developers and excluded groups, but the interests of the agents we analyzed above were paramount in shaping the neighborhood. Within this configuration of the urban space, what process and spatial form does our research object take on?

Spatial processes are the focus, as it is from these that we will be able to understand the geographical process that underpins the formation of Praga Santos Dummont as a street full of bars and restaurants. According to Abreu (2006), "it is necessary to examine, at every moment, the interaction that is established between the economic, social and political processes that develop in the city, and the way in which the space is structured" (p XX).

Correa (1989) defines spatial process as "an expression used by

[6] "What is on the outskirts, in the circumjacencies of somewhere" (DICIO, 2013), in the case of the city, the Central area.

geographers to try to account for what happens in space over time" (p. 37), similar to the spatial practices discussed in this research. The author also adds that "spatial processes are the forces through which the movement of transformation of the social structure, the process, takes place spatially, remaking the spatiality of society" (CORREA, 1989, p. 36), elucidating the social aspect that these processes have.

The processes and spatial forms in the urban aspect are: centralization and central areas; decentralization and secondary nuclei; cohesion and specialized areas; segregation and social areas; and inertia and crystallized areas. Once again, we will only focus on the process of cohesion and the formation of specialized areas, as we believe that this process best explains the configuration of Praga Santos Dummont. However, we would like to point out that all the processes and forms are not mutually exclusive, and there may be cases of multiple occurrences in the urban space to be analyzed.

Spatial cohesion could be defined as "the movement that leads activities to locate themselves together" (CORREA, 1989, p. 56). The activities in this process that help us to understand the agglomeration that took place in our research object are verified by the fact that, although they have no connection with each other in the direct relationship of commerce, they form a functional whole that creates a spatial monopoly, attracting consumers; they are complementary to each other, as they have the same business base and consumer market; and they require personal contacts, preceding greater accessibility.

We can then define Baixo Gavea as an area specialized in the grouping of bars and nightlife services, forming a kind of "entertainment and leisure district", which converges into a complex spatial form of great expression in the neighbourhood and its surroundings, increasing individual circulation through this space. But for geographical science, what is the concept of leisure?

We are therefore going to analyze one of the possible functions of our research object, leisure. We have incorporated concepts from Tourism Geography to provide a more precise epistemological analysis of the subject. Talking about the materialization of leisure in space is, in short, talking about the culture and way of life of the urban population in contemporary society.

Rodrigues (2000) elucidates the difference in the terminology of "leisure" and "leisure":

> The word "leisure" is often confused with the expressions "free time" and "idleness" - these are understood as times that are opposed to working time. [...] The word "leisure", on the other hand, has the connotation of activities, i.e. actions carried out during free time (p. 89).

As such, we are talking about activities that take place during free time in a given urban space, leisure activities that take place during leisure time and do not necessarily represent a tourist activity, despite the presence of several tourists on the site. It is important to emphasize that free time is also used for social obligations and is not only counted as leisure time.

It should be noted that leisure is seen as an activity that brings individual satisfaction and pleasure, as G aelzer (1979) states, "it is the individual harmony between attitude, integral development and the availability of oneself. It is an active mental state associated with a situation of freedom, ability and pleasure" (p. 54).

As we have seen, leisure will correspond to activities that provide pleasure and are carried out freely and spontaneously, regardless of their nature (physical, intellectual, artistic, etc.). These activities are associated with moments of rest, entertainment and amusement, but always comprising an activity and not complete idleness, understood as the act of doing nothing during leisure time.

Castelli (2001) also points out three functions of leisure: (I) rest, (II) fun and (III) development. The first function is seen as repairing the physical and nervous wear and tear caused by the stresses of everyday obligations. The second function can be evaluated as a break with the rhythm of contemporary life. The third function can be seen as developing the personality through participation in freely chosen activities.

When reading up on the subject of leisure, we came across an obstacle related to thinking about the supply of services and the "right to leisure". Garcia (2007) states that free time is a supply of services and products limited by the laws of the market. However, by talking about a limitation of leisure, imposed by a capitalist logic, we would be disregarding the fact that our object is a public space that contains commercial activities linked to leisure. And can a public space contain "private practices"?

## 1.2 The Public Space

To understand the qualification of public space for Geography, we must first think about the main problems related to its definition and academic discussion. The first point is to forget popular notions and, from now on, no longer think of the public as the opposite of the private. This thinking will not allow us to analyze the space in its real sense, limiting it only to the conditions of this conceptual imprecision.

Another problem in understanding the public space is to think only of the legally delimited area, which would lead us to the conception that the law defines the object, which, it turns out, predates any predisposition of codes. Gomes (2012), discussing this problem, says that

> there are various processes in which public activity takes place in spaces that are not strictly and legally public, as in the case of Parisian cafés or English pubs (the name "pub", by the way, comes from the expression "public houses"), so often used as a classic example of public visibility in the 19th century (p. 22).

The bars in Rio de Janeiro, especially in Baixo Gavea, have some of the same characteristics as the pubs and cafés mentioned above, but although they operate in a physically public space (in this case, the Santos Dummont plaza), they are not, by law, public, which would erroneously eliminate a possible definition of a public space.

Finally, a last problem that can be identified in relation to public space is related to its open-access nature, confused by many as the only element to define the public quality of space.

This definition is not borne out in reality because areas such as schools, hospitals and military zones are public and, for the most part, do not allow free access to the population, so we would have the de-characterization of various spaces just because of this opposition.

Gomes (2012) explains that public space, for geography, will have two types of understanding: one referring to concrete physical space (streets, etc.); and, an abstract space that will bring together the foundations of democracy and politics.

Although they are not two integrated spaces, Geography will present a dialog between them, and our understanding of geographical space aims to take into account not only the forms but also the practices and processes that take place. According to Gomes (2012), "the physical arrangement of things and phenomena constitutes an active agent in the realization and qualification of social supports and this spatial order must be perceived as a condition for the production of these supports" (p. 21).

What makes a space public is its relationship with public life, and all the attributes that emerge will also be related to the life of the public man. When we talk about public life, we are referring to "the organization of the common life of men who define themselves as individuals, differentiated and rational, who aim to coexist and benefit from a common life without having to give up their freedom" (GOMES, 2012, p. 24).

The public space will also be the place for dialog, for communication between different individuals, where speeches take place and become public.

Hence an important characteristic, that of the co-presence of individuals in the space.

Another important characteristic of these spaces is that the activities of individuals become public from the moment they are able to contribute to a public opinion through the use of their reason in a debate. Thus, the public space is "the place where problems present themselves, take shape, gain a public dimension and, at the same time, are resolved" (GOMES, 2010, p. 160). This visibility and recognition of problems occurring in a public way are other fundamental characteristics of a public space.

This space also establishes the rules and institutions to which the public will be subject. It is in the space itself that we will define the rules considered "reasonable" for use and transit in it, thus revealing regulated accessibility and isonomy as other characteristics of the space.

The physical spaces are not left out of our analysis, because it is from them that we will configure the behaviors, dispositions and conducts that will follow in the public areas. These modes of behavior will also define the space as public, bringing about a spatial relationship of form and content.

However, we also need to consider the statements made by Gomes (2013), when in his book *O lugar do olhar (The place of the gaze),* he verifies, through experiences, important characteristics about public spaces and the imagination of the individual who will use them. He elucidates the multiplicity of aspects and intentions that an image can bring when it is captured or observed by someone. These aspects and intentions, which are not always positive, can also lead to the consumption or affection of such a space, without there necessarily being any contact with it.

The ideal of the public space is to be the locus that will create the cohabitation of alterities, where diverse collective identities will seek recognition. The public space will form a civic contract, where public men will act with civility, in the sense of carrying out daily exchanges with the different people who share their space.

We agree with Gomes (2012) when he says that we can see that cities always express themselves through a complex network of valorization of their public spaces. The great places where urban life is mixed and staged favor encounters and communication and produce a kind of summary of the socio-spatial diversity of the population (p. 26).

Based on these statements, we consider the group of bars located along Santos Dummont Square to be a public scene, where individuals coexist, where problems gain visibility and recognition, and where there is equality with accessibility regulated by the different venues.

The analysis of the epistemological scope of Geography will serve as a basis for the geographicization of the research object. We will now present the Gavea neighborhood and its functional and geographical complexity.

# CHAPTER 3 - GAVEA AND ITS SPATIAL DYNAMICS

The Gavea neighborhood is an interesting object of geographical research due to its contrasts and rich dynamics, especially its functional heterogeneity over the years. Before we talk about Gavea, we need to define what we are going to talk about. The question *which Gavea*[7] ? has a less intuitive answer than you might think. The name comes from the famous Pedra da Gavea, one of the city's most famous rock formations, which has been in the city's annals since 1502 when "the navigator Gaspar de Lemos cited it among the landmarks used to locate Guanabara Bay for those arriving in Rio de Janeiro by sea" (CORREA, 1997, p. 42).

Until the process of land appreciation in the neighborhood (which took place around the 1950s), the working class lived next to the middle and upper classes, with large "noble" properties next to the factories and laboratories of the industries that occupied the area. The process of valorization will take place with the expulsion of the industries and, later, the replacement of the entire working class population by the middle and upper classes, in addition to the construction of the PUC, the Shopping da Gavea and the Clinica Sao Vicente, milestones of a refunctionalization of the neighborhood.

In order to understand the patterns and flows present in the neighborhood today, we need to analyze its urban evolution and pay attention to its old functions and forms. The history of Gavea is of the utmost importance to the city, as it is one of the South Zone neighborhoods that has undergone the greatest number of changes, following the growth dynamics of Rio de Janeiro. Therefore, we will first analyze Gavea within the history of the

[7] The name *Gavea* refers to the shape of the stone, which was also "named" by the French (La Hune) and the Indians (Piragua). The Aurelio dictionary also defines Gavea as - A *tray set on a wheel of boards at the top of a ship's masts and used for various purposes. / A square sail that occupies the place immediately above the mainsail.*

city, so that we can then approach our object in its entirety.

We would like to point out that it is not in our interest to analyze the city's evolution since the Portuguese conquest of Antarctica, or even the colonial phase; we will focus here on pre-republic Rio de Janeiro, with forms similar to those we find today.

## 3.1 The formation of the parish of Gavea

When we talk about Gavea, we are talking about the southern part of the city of Rio de Janeiro, specifically one of the last neighborhoods to be occupied. The separation by zones in the city occurred due to "the introduction of the donkey streetcar and the steam train which, from 1870 onwards, were major drivers of the city's physical growth" (ABREU, 2006, p. 28). The year 1870 is marked as fundamental for the urban evolution of Rio de Janeiro, as the two fundamental elements of the city's expansion (streetcar and train) began to act in sync.

The different uses of Rio's urban space take different directions, qualifying the city's zones. The "dirty" uses and the less privileged classes will locate themselves in neighborhoods along the train line, while the "noble" uses and classes choose to live around the streetcars. The city will then take shape through its means of transportation, while in the North we can find long, straight streets parallel to the train lines, followed by streets perpendicular to them, in the South the streets form sinuously according to the old paths taken by the streetcars.

The current neighborhoods in the southern zone were derived from chacaras, farm divisions near the urban parishes, which were "reserved for the weekend activities of the ruling classes, and gradually became permanent residences, even justifying the creation of new parishes" (ABREU, 2006, p. 29).

From 1850 onwards, the parishes shown above began to undergo a process of expansion and intensification of occupation, especially the parish of Lagoa, as a result of the expansion that took place in the South Zone from 1938 onwards. It should also be noted that, despite the occupation, the "parishes of the city changed little in terms of their form and appearance between 1838 and 1870, continuing to have an exclusively rural character" (ABREU, 2006, p. 35).

The question of transportation goes beyond just a means of getting around; in our city, transportation also helped to expand the urban fabric and differentiate occupation, with streetcars and trains standing out as distinct qualifications of the people of Rio. While the trains were used to occupy the suburban parishes, which they crossed, the donkey streetcar lines (introduced in 1868) facilitated the expansion of the South Zone of the city of Rio de Janeiro.

On the influence of streetcars on the occupation of Rio de Janeiro, Abreu
(2006) points out

> Not only did the streetcars meet an existing demand, but in meeting this demand, they began to have a direct influence, not only on the pattern of occupation of a large part of the city, but also on the pattern of accumulation of the capital that circulated there, both national and foreign. national capital, coming from a large part of the profits of the coffee aristocracy, merchants and financiers, increasingly began to be invested in real estate in the areas served by the streetcar lines (p. 36).

The owners of this accumulated capital decided on the areas that would be served by streetcars and the provision of urban infrastructure in the city, ensuring that their capital accumulated more and more. New neighborhoods were created and a dissociation between the North and South Zones became increasingly evident.

The role of the streetcars, however, went beyond the accumulation of capital, extending to the intensification of the occupation of the parish of Lagoa, which, being the furthest from the center among the parishes of the South Zone, was not the main target for housing.

Not only have the neighborhoods of the South Zone changed, but the neighborhoods of the North Zone near the streetcar stops have also undergone abrupt changes and noble construction, which can be somewhat similar to the effect of line 2 of the Rio de Janeiro metro today. Luxury condominiums, similar to those built in the neighborhoods of Barra da Tijuca and its surroundings, are now being built in neighborhoods considered to be of low purchasing power, such as Pavuna, Del Castilho and Maria da Grapa, just because of the presence of this means of transport, which used to be a "privilege" for residents of the South Zone.

Returning to the first processes of urbanization and occupation of Gavea, we look at another important factor in Rio de Janeiro: industrialization. The parish of Lagoa, when it was created (in 1795), was designated as an area for the construction of steel factories in its early years:

> SAO JOAO DA LAGOA - Ordering the decree of June 13, 1808, to the Council of Finance, that the Mill and Lands located in Lagoa Rodrigo de Freitas, [...] for the establishment of a Gunpowder Factory, and all others that were necessary for the founding of artillery pieces and rifle barrels [...] be incorporated into the property of the Royal Crown [...] (COSTA, 1963, p. 7).

In 1831, a huge explosion destroyed most of the gunpowder factory, leaving little to be made of the facilities. The war factories that had been installed there were moved to the Serrana region, and workshops sprang up where they had been. From then on, pharmaceutical factories and laboratories, textile industries were born, up to the notorious Companhia de Tecidos Carioca and Fabrica de Tecidos Corcovado, which emerged in the 1880s.

With the increase in the number of factories and the end of slavery, a large influx of immigrants came to work in these industries, causing a process of population growth, aggravating the housing situation in the city, which had not yet undergone any sanitary or urban reform. The government then began to give tax exemptions and benefits to companies that created "hygienic workers' villages, with cesspools, kitchen and laundry facilities, elevated off the ground and with good aeration" (ABREU, 2006, p. 52).

Such a measure would not only help the city's housing situation, but also reduce the cost of the industries in transporting their workers, as well as controlling their actions outside the industry. Government advantages and exemptions led industries in the South Zone to adopt workers' villages, even though the land in the neighborhoods in question was appreciating in value. In 1950, of the 43,000 inhabitants of Gavea, 8,000 were proletarians, forming what many will call *Gavea Vermelha*[8] .

It is worth noting that on June 18, 1873, the parish of Lagoa was dismembered and finally the parish of Gavea was created (Map 1), "whose boundaries stretched from Piagava and Arpoador to Barra da Tijuca" (COSTA, 1963, p. 12). 12). In the 1906 census, this parish had 12,570 inhabitants, of whom 3459 were foreigners, 3569 factory workers and 5589 workers in general[9] , almost 50% of the neighborhood's inhabitants.

---

[8] Allusion to the growing Communism of the Union of Soviet Socialist Republics, whose flag was red.

[9] Source: COSTA, Cassio. Gavea A history of the suburbs. 1963.

**MAP 1 -** The parishes of Rio de Janeiro in the 19th century

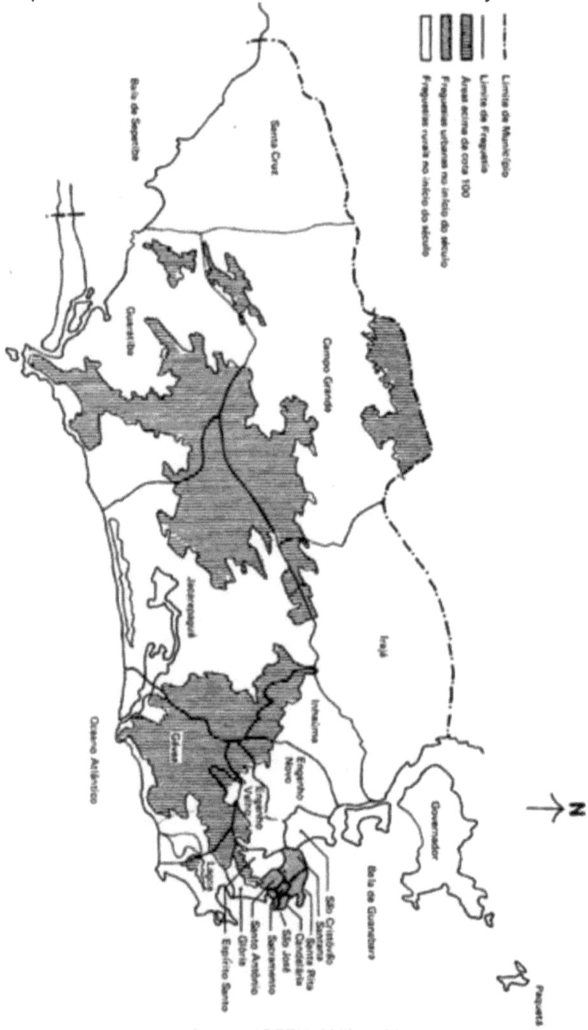

Source: ABREU, 2006, p. 38

## 3.1 The process of upgrading the neighborhood

The beginning of the 20th century was, for Rio de Janeiro, the break with the rudimentary aspects of the city and the beginning of urbanization proper. Mayor Pereira Passos, wanting to accompany the refining of occupation in the South Zone, decided to modify the city center as a whole, but his reforms did not extend to our site studied, limiting only partial sanitation works to Lagoa, as there were still many workers' villages nearby.

At this time, "the governments of the Union and the Federal District, representing the dominant classes, acted preferentially in the sphere of consumption, encouraging the continuation of the process of urban renewal in the central area and the beautification of the southern zone" (ABREU, 2006, p. 68), demonstrating the disparity in the action and development of the city's zones, and, finally, acting in urbanization.

The administration of Mayor Carlos Sampaio (1920-22) was fundamental to the occupation of Gavea and Lagoa Rodrigo de Freitas, and one of his acts was to grant land where the Jockey Clube Brasileiro now stands, opposite Prapa Santos Dummont. The works, which had already begun during the Pereira Passos era, served the mayor as a

> a question of the sanitation of a suburb of our capital, which will undoubtedly be one of the most beautiful and important in the near future. But if the issue of sanitation had to take precedence, the aesthetic aspect was not to be neglected either, especially when Nature was already providing a picture of incomparable charm; And, on the other hand, it was not difficult to foresee that it was perhaps not an advantageous operation from a financial point of view, but at least an undertaking in which the expenses would be entirely covered by the immediate advantages resulting from the sale of the land, without, consequently, taking into account the indirect profits from new taxes and the appreciation of capital (SAMPAIO apud ABREU, 2006, p. 73). 73).

The sanitation works would cause the neighborhood to lose its proletarian character and begin to receive residents with greater purchasing power, acquiring elitist features. The Agache Plan (1930) will justify the presence of the working class population in the South Zone:

> The weaving factories that still exist in Laranjeiras and in the vicinity of the Botanical Garden explain the presence of working-class groups in the districts of Lagoa, Gloria and Gavea, which are, however, markedly residential and aristocratic districts, as can easily be seen from the high percentages referring to domestic service on the one hand, and the liberal and public administration professions on the other, and, finally, the large number of people who live off their income (ABREU, 2006, p. 78). 78).

Also according to the Plan, the Gavea neighborhood would be one of those destined for the wealthy classes, who would be transformed into a garden city of sports, which ended up taking shape around the Rodrigo de Freitas Lagoon. These wealthy classes would later put pressure on the government to expel the industrial activity that still existed there through zoning laws. The construction industry took advantage of the status generated by the ideology of "living by the sea" (now associated with "living in the South Zone") to subdivide and intensify its development.

occupation of the South Zone until the 1950s.[10]

Correa (1997) alluded to the effects of the changes in the subdivision and, above all, the government's concern with making any factor that hindered real estate speculation illegal when he explained that

> In 1937, decree 6.000, which defined the zoning of the entire city, excluded the South Zone from the industrial area. The appreciation of the land led to the rest of the factories being removed until, in 1974, a law was passed that placed almost the whole of Gavea in the ZR-1, a single-family residential zone, which does not prevent, but limits to a great extent, the multiplication of apartment buildings and commercial buildings (p. 50).

Another factor that led to the intensification of the occupation of the wealthier classes in Gavea was when, after 1950, real estate companies, benefiting from the increase in the height of buildings in Copacabana according to their location, decided to build apartments and commercial buildings in order to take advantage of the space and make more profit from it. This, together with the expansion of

Leblon, caused the floating population to increase and the wealthy classes to

[10] It was at this time that the verticalization of Copacabana took place, as well as an unusual growth in the city's service sector. For more information, read GEIGER, Pedro Pinchas. **Evolugao da Rede Urbana Brasileira**, 1963.

prefer the oceanic neighborhoods (ABREU, 2006).

Since then, the changes made in the Zona Sul have been more in the direction of improving traffic, with cars becoming the most used vehicle by the middle and upper classes. The emergence of several tunnels (such as the Reboupas, for example) and the duplication, extension and construction of important avenues are due to this new concern of the government, as well as the end of streetcar lines, the first vector of occupation in the South Zone.

The period of the Military Dictatorship (post-1964) saw two important developments in the attempt to maintain income concentration in the South Zone. The removal of favelas near green areas and any others that violated the strict system of civil rights imposed by the military, and the expansion of the urban core into a new speculative pole, through the construction of the Lagoa-Barra highway, to the neighborhoods of Sao Conrado and Barra da Tijuca; considered by many to be the "brand new South Zone", due to the efforts and benefits given to those residents, reminiscent of the beginning of stratification and differentiation in the city.

As for the proletarians, former workers and residents of Gavea, as their workplaces closed, they moved in large numbers to Rocinha, which ended up also receiving residents of the favelas removed from the South Zone, grew to the point of becoming a reference and, in the 1990s, achieved its emancipation as a neighborhood and administrative region, reflecting the shrinking of the territorial perimeter of the neighborhood in question (Figure 1).

**Figure 1 -** Orthophoto of the Gavea neighborhood today

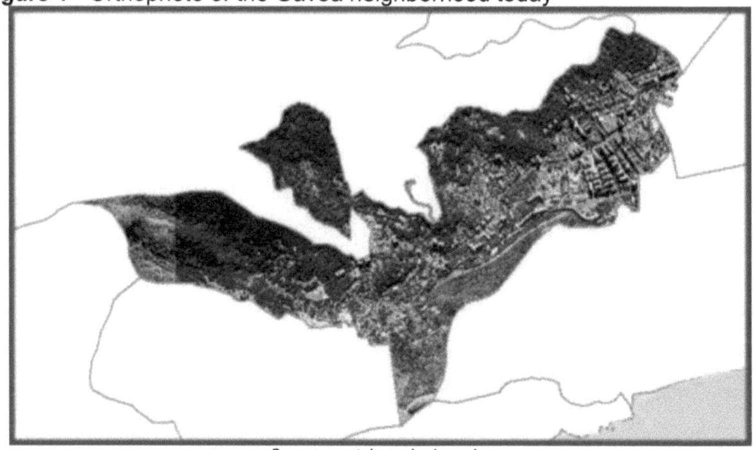

Source: portalgeo.rio.rj.gov.br

Despite having undergone so many changes, not only in urban terms but also in terms of housing, Gavea continued to retain the characteristic that made D. João invest so much in a garden near the industrial area: its natural bucolic amenity. In the words of the poet Carlos Drummond de Andrade:

> to live in Gavea, in one of those streets that go up through nature and seem to end in the sky; streets with resting trees, beautifully singing birds, huge dizzy blue butterflies, the smell of resin, oxygen in pots... And this Rio de Janeiro isn't completely over, as some say. It still has neighborhoods that are blissful.

Gavea's mutability is still present, with many of its forms and functions spread throughout its territorial perimeter. This multidiversity of the neighborhood will also be reflected in one of its main landmarks, Santos Dumont Square, known as *Baixo Gavea.*

# CHAPTER 4 - BAIXO, THE BOHEMIA OF A NEIGHBORHOOD

After discussing the methodological scope that served as the basis for our interpretation, and the process of urban evolution of the Gavea neighborhood that will be the stage for our main subject, we now aim to analyze our research section - Baixo Gavea.

Called by some the only place that can attest to the fact that there is life after dark in the neighborhood, this is the Gavea of bars, restaurants, late nights and meetings with friends. Baixo Gavea is roughly limited to the group of establishments located around Praga Santos Dumont, which is also part of Baixo.

The Lower Town fits the definition of *modern public places,* provided by architecture, because "they are not currently used either for the great popular festivals or for everyday life, and their only reason for being is to [...] break up the monotony of the ocean of houses" (SITTE apud CHOAY, 1979).

This definition, combined with the concept of *public spaces,* can present us with a place that doesn't match the characteristics of a residential neighborhood (despite its apparent location), but which is the locus of public activities, speeches, conversations, entertainment, etc., forming an ambience for those who frequent it.

In this chapter, we will try to understand how this street was formed and its previous function, why the Baixo is referred to as such, its current functional plurality and how the appreciation of the actions of the proletariat, once neglected, inspire the wealthier classes in their leisure. At no point does it seek to end the questioning in itself.

**MAP 2** - Locating Baixo Gavea

Source: portalgeo.rio.rj.gov.br

## 4.1 Largo das Tres Vendas, the predecessor

Before being known as Praps Santos Dumont, and after receiving part of the landfill intended for the creation of the Brazilian Jockey Club, Largo das Tres Vendas (Figure 2) appeared. The nickname was widely used at the end of the 19th century because three famous grocery stores operated here, two facing each other on Rua Marques de Sao Vicente. One of them was recently knocked down to build a condominium of buildings and the other still stands today, where the "Drogarias Cristal" operate.

**Figure 2 -** Largo das Tres Vendas, 1926

Source: Comunidade Jardim Botanico - www.facebook.com/jbcomunidade, Accessed on: 03/07/2013

The Largo grew rapidly thanks to two specific factors: the first was the large concentration of workers and proletarians in general who visited and shopped in the grocery stores located there, and the second was the fact that it was designated as the end point of the Gavea streetcars. Although there aren't many historical records, we believe that this was Praga's first moment as a meeting place for public discourse. The intense flow of people also attracted nearby farmers who saw the open-air markets as another opportunity to earn income for their families.

Another important point in the Largo was the *Hotel do Amaral,* owned by Captain Manuel dos Anjos Victorino, who paid for the construction of the Nossa Senhora da Conceigao parish church in 1857, which, due to its beauty, ended up generating a new nickname for the plague. This nickname was eventually superseded by tributes to the philanthropist Ferreira Viana and President Artur Bernardes, until in 1932, to honor the death of the father of aviation Santos Dumont, the government of the Federal District gave his name to the plague (COSTA, 1963).

Let's stop talking about Lower Gavea for a moment to introduce its contrast, Upper Gavea. It was common in the parishes for the nobles and the wealthier classes to live on the higher topographies, seeking to get closer to the kings and emperors, who used this strategy as a symbolic reaffirmation of power and the possibility of watching over their entire kingdom (OLIVEIRA; KUHN, 2011).

Despite the lack of transportation and the distance from the city center, the neighborhood was sought after by the illustrious as a permanent or summer residence due to the latent presence of the elements of nature there. Among the residents of the upper parts of Gavea we can mention Almirante Marques de Tamandare, Barao do Lavradio, Doutor Osvaldo Cruz, Ferreira Viana, Conselheiro Lafaiete Muller, among many other barons, viscounts and high-income earners.

Even after the proclamation of the Republic, the houses in the higher areas of Gavea never ceased to be occupied by wealthy families, and today there are still several clubs and large mansions along the streets that cross Marques de Sao Vicente.

There was also a *Medio Gavea,* when the term was not used

pejoratively and referred only to the topography of the neighborhood. *Medio Gavea* was located where today we have the Pontifical Catholic University - PUC - and its surroundings. This university used to be a pottery factory, with many proletarian residents around it.

## 4.2 *Baixo Gavea* with "old habits" and a "renewed" public

The change in the concept of the word "*low*" is reflected in two of its definitions in the dictionary. There is the definition that was once used, not pejoratively, where it referred to the "lower part: the bottom of the mountain", which expressed the representation of the location in relation to the whole parish, however, along with the plans for the elitization of the South Zone and its neighborhoods, came the beginning of the use of low as "Inferior, despicable, ungraded: the lower class", to diminish those who lived there, mostly proletarians.

It's worth noting that, despite being practically next door to *Baixo Gavea,* the Jockey Club never suffered any kind of reprisal during the period when the factories and proletariat were expelled from Gavea. To this day, equestrianism is not considered a popular sport, which means that the territory established there by the regulars was free of any "low" pressures.

One of the notions worked out by Piaget in his Basic Cognitive Theory is that any child, in the concrete operant stage, can already differentiate and define the big things from the small, the high from the low, but this same child ends up incorporating into their memetic base that the biggest and highest will always be the best places to be (SOUZA; KRAMER, 1991). Could we then think that, regardless of the other ways of defining *low,* no one could ever think of it as something positive?

The truth is that we are led to believe from birth that *low* is a bad thing, expressions like "being below", "staying below", "lowering your head" and so many others show us how bad it can be to be *low*. However, as soon as they managed to "expel the unwanted classes", there was no longer any kind of prejudice against the lower class, because it would be refunctionalized with bars, restaurants, warehouses and so many other establishments.

While new businesses were being set up around Praga Santos Dumont, the Praga itself had not changed. On the contrary, the project of constant beautification of the South Zone, still latent in the post-1950s, ensured that Praga remained a widely usable public space.

The change in landmarks was not complete, instantaneous or modernizing. Today we can still see many different types of buildings in the streets that cross the Baixo, as Maleque (2007) points out: "old houses from the beginning of the century, [...] multi-family buildings built in the 1940s and 1950s, as well as the large frame buildings that emerged in 1976" (p. 115).

And although it has a noble air about it, *Baixo Gavea* has consolidated itself over the years as a bohemian place, and one of the exponents of the welcoming identity of the carioca, through the process of spatial cohesion of the bars and restaurants. Nowadays, it's a place where you can find people dressed in the most varied ways, casual, formal, ready to go out at night, etc.

The most interesting thing to observe is how Praga Santos Dumont undergoes a spatial transformation in a few minutes, being completely reconfigured for the well-being of the bohemians who will be there.

It's worth noting that, despite being included on the city's tourist maps, *Baixo Gavea* has suffered a severe blow to its routine. In December 2010, Mayor Eduardo Paes, using a decree based on the Municipal Organic Law, instituted the Silence Law, requiring establishments to close from one o'clock in the morning, with permission to close at three o'clock in the morning on Saturdays and Sundays.

It's not difficult to find this area in the main media outlets, either because of the presence of actors, musicians of the most diverse styles, or because of the musical references, even if they are often subtle, or simply because of the entertainment tips. It is, as a rule, a good place to visit in the city of Rio de Janeiro, as observed in field research and informal interviews.

*Baixo Gavea,* with the bonus of being part of a public space, also becomes plural and multi-purpose. Praga Santos Dumont, with its children's playground, benches and paths cutting through the well-kept gardens, becomes a picturesque setting for strolls and casual encounters.

We could only go into the aspects of Prague, but we would like to highlight two activities that take place with different frequencies, but which mark the *Lower* Town. One activity would be the open-air and antiques fairs, which take place every Friday and Sunday, respectively; and the other is the carnival blocks that liven up the neighborhood's streets for almost the entire week of Carnival.

Jesus (2005) explains free markets as

> an initiative taken by Pereira Passos, which, together with other measures regarding the use and formatting of the city's public spaces, would form part of a broad project for the pedagogy of civilization. In these diverse territories (parks, avenues), some of which were ephemeral/periodic (open-air fairs, the battle of the flowers, etc.), citizens would experience the triumph of the modern city, re-educating their behaviour (dress, grooming, sociability) and rethinking the society in which they lived, seeing it as progressive and promising, with great potential for redemptive civilization.

Considering street markets as a model of civilizational re-education is valid, but it's not the only way to think about them. When we talk about the importance of the observer's experience in shaping the public space, we have to take into account that this already re-educated (or educated) citizen can find in this locus a familiar and entertaining ground beyond just something organized. When, for example, an individual buys something to eat at the market and, instead of stopping and sitting down, decides to eat while walking, is he transgressing the pedagogy of civilization or is he just acting normally and informally?

Like Gomes (2012), we think of the open-air fairs (and of course the carnival blocks, as well as the bohemian hours of *Baixo)* as a way of *doing things in the* public space. An action that is not free from social and governmental rules, but which obeys nothing but its own desires and conventions.

The other activity takes place only during Carnival, and also brings together the most varied public in the name of popular entertainment. As I observed during my fieldwork, there are several blocos that belong to Praga Santos Dumont, and many that take advantage of it to start their routes. During this time, the bars and restaurants coexist with the lively blocos, bringing a third and specific spatial configuration to *Baixo Gavea,* as we can see in Figure 3, respectively Praga Santos Dumont, the free market, the carnival blocos and the night in *Baixo Gavea.*

**Figure 3 -** The four periods of *Baixo Gavea*

Sources: riofilmcommission.rj.gov.br; f eiramania.com.br;
Personal archive, 2012; leconcierge.uol.com.br; Accessed on: 04/07/2013.

The success of Baixo Gavea in bringing together establishments that promote nightlife in a residential neighborhood has led to several other entrepreneurs taking the liberty of building Baixos around the city.

Almost always achieved through the process of cohesion, the Baixos are formed and guarantee the nightlife of neighborhoods that can be considered calm or peaceful to live in. This is how *Baixo Meier* (located near the train station), *Baixo Tijuca* (located on Praqa Varnhagen), *Baixo Leblon* (located between Dias Ferreira and Aristides Espinola streets and Avenida Ataulfo de Paiva), among many other *Baixos,* came about.

If we could come up with a simple definition of what a *Bass* would be, we could think of it in the sense of spatially cohesive areas that excel at night-time entertainment, making use of public spaces within neighborhoods where most of the buildings have residential functions.

# CHAPTER 5 - FINAL CONSIDERATIONS

The monographic work presented on *Baixo Gavea* sought to initiate an academic discussion about a term that is popularly used, but has no apparent scientific definition. By presenting and attempting to provide a theoretical basis for this term, we hope to contribute to Geography in order to conceive, in the future, a broader debate on this object. We would like to point out that our small amount of reading and incipient scientific participation do not yet allow us to discuss the subject in depth.

In this sense, we searched for readings that would help us with urban issues and bring greater scientificity to the object, as well as carrying out fieldwork. The attempt to take a common-sense vocabulary and turn it into something minimally scientific stimulated our search, but the lack and/or scarcity of specific material on the subject was one of the major obstacles we encountered in our research.

In this way, in an attempt to understand the spatial practices that take place and all the agents involved, we sought to specify our methodological scope with urban issues and the concepts adjacent to them. Understanding the spatial processes and forms stimulated by the various social agents present makes us reflect directly on the functions they can fulfill in the production of space.

By analyzing the object on the larger scale of the city of Rio de Janeiro, and narrowing down this same analytical scale until we reach Praga Santos Dumont, not forgetting to pass through the South Zone and the Gavea neighborhood to which it belongs, we become involved in a geo-history of one of the city's most changing neighborhoods.

The evolution of the South Zone of Rio de Janeiro is easily understood by the state interventions necessary for its stratification and differentiation from the North Zone, but Gavea, being the parish furthest from the Central Area, and a neighborhood with numerous factories, did not receive such considerations. When the real estate potential of the South Zone was noticed, and there was also a need for the wealthy residents of Copacabana to "flee", the state stepped in and helped the industries present so that they could leave as soon as possible, looking for new, less valuable spaces in the city.

Thus, Gavea developed into an elite neighborhood, with one of the "remnants" left by the proletarian population as a defining characteristic: the bars and meeting places of *Baixo Gavea.* The neighborhood's residents readily embraced this trend, which has continued to this day and is well-liked by all those who frequent it.

Discussing the dynamics and plurality of our object of research, we saw that *Baixo Gavea* is not just about night-time activities, but also has three other possible ways of using public space: the public plague; the free markets; and the stage for carnival blocks. All of the forms presented and discussed in the paper attempt to briefly explain how the objectivity of man's use of space can happen at different times, in different ways, expressing, almost in full, all the possibilities of spatial practices that can occur, configuring different social environments.

Finally, we would like to emphasize how difficult it is to consider and start debates on a spatial configuration that has not yet been specifically considered by the academic community. We believe, within our possibilities, that we have elucidated and discussed the proposed issues in such a way as to serve as a starting point for future academic research and analysis.

With all this in mind, we also believe that we have made a significant contribution to the geography of Brazil, especially Rio de Janeiro. In this way, we believe that geographic science in Brazil should try to carry out more research that appropriates popular culture or common sense, and make it part of our daily studies. Breaking the barriers imposed by other human sciences and always being at the forefront of academia, even on a national stage, because we have brilliant academics and still some subjects that, however simple they may seem at first, are not so well explained.

We would like to outline possible ways forward, in order to improve what has been presented here and develop the research along other lines. One of the points we could discuss specifically is the creation and incorporation of a genuinely carioca identity, after all, there are a number of articles from various sciences using the expression "carioca identity" without it being properly discussed. Another point that can be brought up is the specific research into the current and previous establishments in *Baixo Gavea,* trying to elucidate their interests and possible particular territories/spatial practices that led them to success/failure. A final point that also interested us was the relationship between the residents of the surrounding area, regardless of their historical time and social class, and the different temporalities in which *Baixo Gavea* lived. Their ways of acting in the space, their experiences and their perspectives on this public space would also be a great advance on the subject.

# REFERENCES


ABREU, Mauricio. **Urban Evolution of Rio de Janeiro**. Rio de Janeiro: Pereira Passos Institute, 2006.

ANDREATTA, Verena; CHIAVARI, Maria Pace; REGO, Helena. O Rio de Janeiro e a sua orla: historia, projetos e identidade carioca, **Colegao Estudos Cariocas**, Rio de Janeiro, n. 20091201, dec. 2009. Available at: <http://portalgeo.rio.rj.gov.br/estudoscariocas/download/2418_O%20Rio%20de%20Janeiro%20e%20sua%20orla.pdf>. Accessed on: February 6, 2013. 16p.

BESSA, Kelly. Diferenciagao espacial como elemento proprio a natureza da Geografia, **Mercator**, Fortaleza, v. 9, n. 20, p. 43-56, set./dez 2010. Available at: <http://www.mercator.ufc.br/index.php/mercator/article/viewFile/394/308>. Accessed on: November 28, 2012.

CASTELLI, Geraldo. **Tourism: a remarkable activity**. Caxias do Sul: EDUCS, 2001.

CAVALLIERI, Fernando; LOPES, Gustavo Peres. Indice de Desenvolvimento Social - IDS: comparando as realidades microurbanas da cidade do Rio de Janeiro, **Colegao Estudos Cariocas**, Rio de Janeiro, n. 20080401, abr. 2008. Available at: <http://portalgeo.rio.rj.gov.br/estudoscariocas/download/2394_%C3%8Dndice%20of%20Development%20Social_IDS.pdf>. Accessed on: February 6, 2013. 12p.

CHOAY, Frangoise. **Urbanism**. Utopias and Realities - An anthology. Sao Paulo: Perspectiva Publishing House, 1979

CORREA, Roberto Lobato. **The Urban Space**. Sao Paulo: Atica, 1989.

CORREA, Marcos Sa. **Gavea**. A history of Rio. Rio de Janeiro: Agir, 1997.

CORREA, Roberto Lobato. Space, a key concept in Geography. In: CASTRO, Ina Elias de; GOMES, Paulo Cesar da Costa; CORREA, Roberto Lobato (Orgs.). **Geography: Concepts and Themes**. Rio de Janeiro: Bertrand Brasil, 2008

COSTA, Cassio. **Gavea**. History of the suburbs. Rio de Janeiro: Department of History and Documentation of the State of Guanabara, 1963.

GAELZER, Lenea. **Leisure: blessing or curse?** Porto Alegre: Federal University of Rio Grande do Sul, 1979

GARCIA, Rita Maria de Paula. Produgao do Espago pelo Lazer e Turismo, **Revista Eletronica da Associagao dos Geografos Brasileiros - Segao Tres Lagoas-MS**, T res Lagoas, v.1, n.° 5, p. 113-126, maio 2007.

GEIGER, Pedro; ARUEIRA, Luis Roberto; ALEM, Adriano. Mapa social da cidade do Rio de Janeiro, **Colegao Estudos Cariocas**, Rio de Janeiro, n. 20010801, Aug. 2001. Available at : <http://portalgeo.rio.rj.gov.br/estudoscariocas/download/2325_Mapa%20social%20da%20cidade%20do%20Rio%20de%20Janeiro.pdf>. Accessed on: February 6, 2013. 27p.

GOMES, Paulo Cesar da Costa. **A condigao urbana: ensaios de geopolitica da cidade**. Rio de Janeiro: Bertrand Brasil, 2010.

GOMES, Paulo Cesar da Costa. Public Spaces: a way of being of the space, a way of being in the space. In: CASTRO, Ina Elias de; GOMES, Paulo Cesar da Costa; CORREA, Roberto Lobato (Orgs.). **Geographical Views**. Ways of seeing and experiencing space. Rio de Janeiro: Bertrand Brasil, 2012.

GOMES, Paulo Cesar da Costa. **The place of the gaze: elements for a geography of visibility**. Rio de Janeiro: Bertrand Brasil, 2013.

JESUS, Gilmar Mascarenhas. Ordenando o espago publico: a criagao das fairs livres na cidade do Rio de Janeiro, **Revista Electronica de Geografia y Ciencias Sociales, Universidad de Barcelona**, v. IX, n. 194(62), August 1, 2005. Available at: <http://www.ub.edu/geocrit/sn/sn-194- 62.htm>. Accessed on: July 2, 2013.

MALEQUE, Miria Roseira. From a proletarian neighborhood to the elegance of Gavea. In: LIMA, Evelyn Furquim Werneck, MALEQUE, Miria Roseira (Orgs.). **Space and the city: concepts and readings**. Rio de Janeiro: 7letras, 2007.

MORAES, Antonio Carlos Robert. **Geografia - Pequena Historia Critica**. Sao Paulo: Annablume, 2003.

OLIVEIRA, Clarissa Wetzel; KUHN, Fabio. Arquitetura como codigo de segregação: considerações sobre os usos da arquitetura e aplicabilidade do conceito de segregação no período colonial, **Cordis**, n. 6, p. 277-300, jan./jun. 2011. Available at : <http://revistas.pucsp.br/index.php/cordis/article/viewFile/10301/7690>. Accessed on: July 2, 2013.

RODRIGUES, Adyr Balastreri. Geography of tourism: new challenges. In: TRIGO, Luiz Gonzaga Godoi (Org.) **Tourism**. How to learn, how to teach. Sao Paulo: SENAC, 2000.

SANTOS, Milton. **The nature of space**. Technique and time, reason and emotion. Sao Paulo: Hucitec, 1996.

SILVA, Edima Aranha. Lazer nos Espagos Urbanos, **Revista Eletronica da Associagao dos Geografos Brasileiros - Segao Tres Lagoas-MS**, Tres Lagoas, v.1, n. 1, p. 54-68, nov. 2004.

SILVA, Rodrigo Kuhn. The evolution of the concept of geographical space. In: **XVI SIMPOSIO DE ENSINO, PESQUISA E EXTENSAO**. Rio Grande do Sul: Unifra, 2012. 9p.

SOUZA, Solagen Jobim; KRAMER, Sonia. O debate piaget/vygotsky e as politicas educacionais, **Cadernos de Pesquisa**, n. 77, p. 69-80, May 1991.

SUERTEGARAY, Dirce Maria Antunes. Espago geografico uno e multiplo, **Revista Electronica de Geografia y Ciencias Sociales, Universidad de Barcelona**, n. 93, 15 de julio de 2001. Available at: <http://www.ub.edu/geocrit/sn-93.htm>. Accessed on: November 28, 2012.

VELHO, Gilberto. The worlds of Copacabana. In: VELHO, Gilberto. **Urban Anthropology: Culture and Society in Brazil and Portugal**. Rio de Janeiro: Zahar, 1999.

MIX
Papier aus verantwortungsvollen Quellen
Paper from responsible sources
FSC® C105338

Printed by Books on Demand GmbH, Norderstedt / Germany